SERGEANT MAJOR POTTS WORLD OF SCIENCE

This book was possible because of

Trent
Shelly
Trevor
Nick
Joey &
Leah

SERGEANT MAJOR POTTS WORLD OF SCIENCE

CONTENTS

About the author ******************************* 3

Introduction to the precise, sacred, scientific procedures **** 4

Sugar rock diamonds and recycle paper into a tree ******** 5

Magic eye dropper and pirates ************************ 7

Flying tinsel and invisible air ************************ 9

Magic sand and gassy every day and not unusual bottle **** 11

Painting flowers ********************************* 13

Plastic milk ************************************* 14

Surface tension ********************************* 15

Dancing spaghetti ******************************* 16

Balloon rocket ********************************** 18

Food facts ************************************* 20

Magic potato *********************************** 21

Magic balls and baggie race *********************** 22

Things to do with balloons ************************ 23

Snake magnets ********************************** 24

Paper magic ************************************ 26

Paper experiments ****************************** 27

Farewell from S. M. Potts ************************* 28

SERGEANT MAJOR POTTS WORLD OF SCIENCE

About the author:

James is a spiritive leader of a motley crew. He sailed from the Belgium Congo to Indochina using only a bushel of plantains and an elephant gun as a paddle. He tried his hand at Formica mining and bear wrestling in the Yukon. He was bold and impetuous. He once held off an entire flotilla of Spanish conquistadors single handedly, only to be nearly swallowed whole by the elusive Kraken. Fortunately, the leviathan has an extreme weakness for Cuban rum and Tootsie Rolls. His personal fears include marmosets and peanut brittle. His dedication for producing this publication was to provide inspiration and motivation of the world of science to youth. Thanks to the World of Wonders Science Museum in Lodi California for providing imagination and motivation of science to all.

Trek on!

SERGEANT MAJOR POTTS WORLD OF SCIENCE

Welcome to the scientific world of Sergeant Major Potts. Potts served in the British Army and later became an archeologist and great explorer. The following precise, sacred scientific procedures were discovered in 1906 by Major Potts in the Egyptian tombs of the pharaohs. His ledgers of exploits were recently found and are now forwarded to you. It is up to you to demonstrate these scientific experiments to others so they may learn the great mysteries of science.

You will bring motivation and curiosity to minds of thousands. Maybe hundreds. Possibly 50. More than likely 4 to 5. Follow these procedures, don't let Sergeant Major Potts down and read the instructions imphatically…emfatically…imfhatically…the best that you can.

SERGEANT MAJOR POTTS WORLD OF SCIENCE

Sgt. Major Potts wonders of science with secrets from the tombs of Egyptian Pharaohs. Did someone say ROCK Candy?

Going through my journal, I ran across the time that I needed to prevent the evil Cornelius Snod from taking rare diamonds from antiquities. I swapped the diamonds for rock candy. He thought that he stole the diamonds but actually took sugar rock candy that looked like diamonds. Snodd was apprehended and the police enjoyed the sugary-sweet rock candy.

What you will need:

- 1 cup of water
- 3 cups of sugar
- Clean glass jar
- Pencil
- # 16 cotton twine

With help from an adult, heat the water in a saucepan over medium-high heat until it comes to a boil. Completely dissolve the sugar in the boiling water. Add more sugar until the water is like syrup. Stir continuously with a wooden spoon until the solution becomes clear. If you want rubies or emeralds this is the time that you can add a few drops of food coloring.

Remove the solution from the heat, and then carefully pour it into the clean jar. Cover the jar with a small piece of aluminum foil to keep out dirt.

Dip the twine into the sugar water and remove. Roll the sticky twine in dry sugar, and allow it to dry (1 to 2 hours). Dipping the twine in sugar acts as the "seed crystal" to encourage new crystals to cling to the string.

Tie a weight, such as a paperclip to one end of the twine, and then tie the other end to the middle of a pencil. Place the twine in the jar so that the weight sits 1/2 inch from the bottom of the jar. Let the jar sit at room temperature, undisturbed, for several days. You can check each day to see how much your crystals have grown. If no crystals are forming, heat in a saucepan again and add more sugar.

What is going on?

The word crystal refers to any matter, such as molecules, that is arranged in an orderly form. Once the solution starts to cool and evaporate, the loose sugar molecules start to join with the sugar molecules

SERGEANT MAJOR POTTS WORLD OF SCIENCE

on the twine. These molecules gradually join with more sugar molecules on the twine and become crystals. You now have tasty diamond sugar crystals.

This is Sgt. Major Potts world of science signing of until next time.

ONE MORE AMAZING SPECTACLE

You will take ordinary newspaper and recycle it into a tree.

1. Ok, take three sheets of newspaper.

2. Tape them end to end with one piece of tape at the bottom, one at the top and one in the middle.

3. Roll the paper up. Rolling around a wrapping paper roll helps, but don't roll up too tight. Then take it out of the wrapping paper roll. Only if ya used one.

4. Tape the outside bottom of the paper roll and also half way up.

5. Squeeze the top of the roll and cut from the top half way down the paper roll. Rotate a quarter turn and make another cut half way down.

6. Next bend back the 4 strips that you cut.

7. Take hold of the inside center and say, (with an authoritative voice), "I will recycle this newspaper back into a tree." And with the magic words "Shezam, Gismo, Neato Jet." You then lift the center of the paper and it will turn into a tree. Yes, it is amazing and possibly scientific.

SERGEANT MAJOR POTTS WORLD OF SCIENCE

Potts journal January 1906. You are now in the midst of the amazing, super-duper, Sergeant Major Potts world of Science and the magical eye dropper.

Potts heard news of a sunken, ancient Asian ship Liki Tiki that had just been discovered due to a combination of the winter monsoon rains, wind, heavy waves, and low ocean tides. "Gad Zookes," Potts expressed. "The looter pirates will find the great treasures before me." Potts took the first China Clipper airplane to the South Asian coast and swam down to the newly revealed sunken ship. Unfortunately, three pirates arrived soon after, capturing Potts with fishy, smelly, sharp harpoons pointed at him. The pirates thought to use Potts as ransom for money, but Potts had other ideas.

Potts told them if he gave them a magic bottle, will they let him go? The pirates agreed. So Potts whipped out an everyday and not unusual plastic bottle and an eye dropper. He placed the eye dropper in the bottle and secured a cap on the top of the bottle. Holding the bottle with one hand, he said, with an authoritative voice, "Shezam, Gismo, Neato-Jet." And the eye dropper magically sank to the bottom of the bottle. The pirates looked at each other in amazement. They grabbed the bottle and left. The treasure was returned to the museum of antiquities the very next day along with a pet tiger named Kumar to guard the treasures against future pilfering. Was the bottle amazing? Yes. Was it magic? No way. It was science that you too can now do.

What you will need:

All that you need is a 2 liter, clear, every day and not unusual plastic soda bottle, an eye dropper, and water.

Fill the bottle all the way to the top with water. Squeeze water into the eye dropper up to the base of the black rubber squeezer thingy. Place the eye dropper into the bottle and screw the lid onto the bottle tight.

Now place the bottle on a table. With one hand hold the bottle and with the other hand, unless you have three, (use your judgment), pretend there is a string attached to the eye dropper. When you move your hand down the outside of the bottle, squeeze the bottle with the other hand. It will look as if you have magical powers moving the eye dropper inside of the bottle. As you squeeze the bottle, you are causing water pressure that is sending more water into the eye dropper, making it heavier and it will fall to the bottom of the bottle. . Some people may go crazy trying to figure this one out. Yes, this is the time that someone may pass out or can't remember their-own name. And don't forget the magic words. If the eye dropper does not go down easily, take it out and add more water.

SERGEANT MAJOR POTTS WORLD OF SCIENCE

ONE MORE AMAZING SPECTACLE

Ok, now we are going to prove that science and chemical reactions can be really weird some times. We will make a bouncing ball using wood glue and soap. I told you it is weird. Since it is not something that you would usually keep in your pocket, or at least I hope not, search for Borax soap and a 4 ounce bottle of white or clear Elmer's glue.

In a container, combine one tablespoon of Borax to two cups of warm water and stir until the Borax is dissolved. To make it real neato-jet, put in a few drops of food coloring. Now squirt the glue into the container of Borax solution and stir the contents. You can now use your hands to knead it like dough. Now tighten your seat belt because magical mystifying molecules are doing weird things unknown to the human eyes. The Borax molecules are changing the glues molecules by eliminating one molecule in the glue so the other glue molecules will stay linked together. This is called polymer cross-link. Did I lose you?

Ok imagine 6 students walking around each other. They are the glue molecules. Now 6 other students, the Borax molecules, come in and grab onto their glue molecule hands producing a chain of one big blob. That is what you just made. Your blob is similar to silly putty that can be made into a ball. And it bounces. Separate it into little blobs and place it into a container. After a few moments, those molecules will all hold hands again and form back into one big blob. It will eventually dry out so keep it in a plastic bag. And Mom, it can wash out with soap and water. Is it magic? No way! It is science.

SERGEANT MAJOR POTTS WORLD OF SCIENCE

Potts journal, March 1906 Cheers to you, future science specialists of the world to the next mystifying wonders of science and the magic of the white baton.

When Sergeant Major Potts was on an Amazon jungle expedition, looking for the next amazing science discovery, his team was surrounded by a poison dart shooting pigmy tribe with poison shooters pointed at them. Sweat was rolling off the team members, who were surprised by the fatal situation in the hot, humid jungle. But Potts whipped out his amazing, every day and not unusual, white plastic baton and pointed it at the pigmy tribes. The baton was so close to their faces, they became cross eyed. Potts then whipped out a piece of wool and silver tinsel. He then said his magic words, "Shezam, gismo, neato-jet" and the silver tinsel floated in midair. The pigmies fell to their knees to praise the amazing every day and not unusual white plastic baton spirits.

So Sergeant Major Potts saved the day and now you can do the same. Just follow these precise directions in case you want to amaze people or are overtaken by Amazon, poison dart pigmies.

What you will need:

One 12 inch, or longer, PVC pipe, wool or silk cloth, and thin strips of wrapping cellophane or Christmas tinsel.

Hold the everyday and not unusual white plastic baton, PVC pipe, with one hand. Take wool or silk cloth with the other hand, (unless you have three), (use your judgment), and rub the cloth up and down the baton at least 15 times.

You will hear crackling sounds making the cloth and baton a negative charge.

Since the baton is negative charged and when you drop the positive charged tinsel on the baton they attract.

As soon as the tinsel touches the baton it picks up a negative charge. Since the baton and tinsel are now negative charged, the repel each other and the tinsel will float above the baton.

Sometimes you need to shake the baton to separate the tinsel.

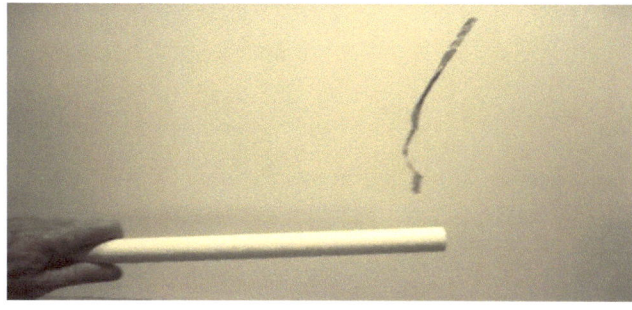

SERGEANT MAJOR POTTS WORLD OF SCIENCE

It takes some practice to get the silver tinsel to float, but if it does, be sure to say the magic words when it does, with an authoritive voice, "(shezam, gismo, neato-jet)". This will amaze and astound people. Make sure no one is chewing bubble gum because they may choke on it when they see such a magical, scientific feat. People may also think you have Harry Potter magic.

WARNING!!! If the tinsel touches anything such as walls, ceiling, faces, peanut butter, the electrons will jump off of the tinsel and you would lose the magic and have to start over or dodge pigmy darts.

ONE MORE AMAZING SPECTICLE

WARNING!!! This experiment can be done **ONLY** with an adult to help. OK???!!!
1. You will need three candles. An adult will light the candles.
2. In a 4 cup, or there abouts, measuring device measure in ½ tablespoon of baking soda.
3. Next pour in ¾ cup of vinegar.
4. You will see the liquid foam up. Do the next procedure quickly.
5. You may want help to pour the measuring device over the candles but don't let the liquid pour out. If all goes well, the candles will mysteriously go out.

What happened? The baking soda is sodium bicarbonate and has chemical energy. The vinegar is an acid and separates the baking soda into carbon dioxide and water. The bubbles you see are carbon dioxide gas that is heavier than air. It flows like water when you pour it. Candlelight needs oxygen to burn but the carbon dioxide removes the oxygen, so the candle goes out. Before pouring, you may need to place paper at the top half of the cup, or thereabouts, to keep the invisible carbon dioxide at the bottom so it may stream out better. That's another amazing scientific demonstration thanks to Sergeant Major Potts World of Science.

SERGEANT MAJOR POTTS WORLD OF SCIENCE

Potts journal, May 1906. Well, the anticipation is over. Here are more Sergeant Major Potts science secrets from the tombs of the Egyptian pharos. The past experiments were amazing beyond belief. But hold onto you hats, your toes, your peanut butter sandwiches, or anything else to keep you from falling over because here are more amazing, beyond belief, science secrets that we are entrusting to you to pass on to the world, or thousands. Maybe 100. Possibly 50. More than likely 10 or 15.

A Sergeant Major Potts quote:
"Aim high, fly straight, and don't crash."

Back when Sergeant Major Potts was exploring through pyramid tombs, he came upon a room with a golden chest that was resting on a dusty table. Potts slowly opened the chest and found white sand inside. He put much of the white sand into a canvas bag because he figured that the sand must have magical powers. Lucky for Potts, he tied the canvas bag onto his belt. When exiting the pyramid tunnel, Potts was greeted by 50 sword swinging Grijari soldiers. They weren't happy that Potts was inside their sacred pyramid. They were angry, and also had bad breath, and were coming closer to Potts with their sharpened blades.
With a snarl and a grit of his teeth, Sergeant Major Potts grabbed a hand full of the magic white sand from his canvas bag and threw it into a puddle of muddy water that was between Potts and the bad-breathed, yellow-teethed, angry soldiers. Potts raised both hands and said the magic words, "Shezam, gismo, neato-jet" and the muddy puddle began to grow higher and higher.

This scared the soldiers so much that they ran away in a cloud of dust. Potts also had a mystified look on his face that turned into a grin. He then said, "Ah". "This is another amazing, shocking, awesome, scientific experiment." Now it is your turn to demonstrate the magic of the white Egyptian sand to others.

SERGEANT MAJOR POTTS WORLD OF SCIENCE

To do this experiment you must order on the computer from Amazon a tube of Insta-Snow. You will also need a larger plate and 2 ounces of water.

Place 1/2 cap full of magic sand in the middle of the saucer. Then pour the 2 ounces of water over the sand and with an authorative voice, you know what to say, "Shezam, gismo, neato-jet" and the sand should expand 100 times its original size. Did I hear someone say Wowie Zowie!!!??? The magic sand is called super absorbent polymer. Polymer is synthetic, plastic molecules that are linked together, as if they are holding hands and not letting go, and expand 100 times their size when absorbing liquid such as the properties of a sponge. When your amazed friends touch the expanded polymer, tell them that this is the stuff that they put in baby diapers to absorb moisture. Ouououou! Let the polymer dry without compressing it and you can use it over again. You will also see that it returns to its 1 ½ capful size after the water is all gone. The super absorbent polymer was originally developed for the forest service that plant seeds in the polymer, since it holds water. The seeds do not dry out and will sprout into trees. Amazing? Yes! Is it magic? NO WAY!!! It is science!!!!!!!!!!!!!!!!!!!!!!!!

ONE MORE AMAZING SPECTACLE

What you will need for this experiment:

One 2 liter EMPTY clear plastic, every day and not unusual soda bottle, cheap cooking oil, food coloring, and an Alka-Seltzer.

Now, pour the cheap cooking oil 4 inches from the bottom of the empty soda bottle. Add water until the contents reaches 2 inches below the top of the bottle. Wow! You will see the water separate from the oil. What is heavier the oil or the water? Now add one or two drops of food coloring, of your choice, in the bottle. Wowie, Zowie, watch the food coloring drop into the oil. "Neato-jet." But that isn't the amazing spectacle. Now break the Alka-Seltzer into 2 pieces. After the oil and water separate, drop a few Alka-Seltzer pieces into the bottle and watch the show. Now that is the amazing part! The carbon dioxide gas rises sending water drops up into the oil. But since water is heavier than oil, the water floats back down. You now have a water and oil Alka-Seltzer lava lamp.
WARNING: The Alka-Seltzer makes a carbon dioxide gas. If you twist the bottle cap on tight, while the gas is bubbling, you will get quite an explosive oil-water mess. [MOM]: after the Alka-Seltzer is finished fizzing, you can then put the cap back on to save the contents, and save from cleaning a mess, because you will be able to do the experiment over and over again.
This is Sergeant Major Potts World of Science signing off until next time.

SERGEANT MAJOR POTTS WORLD OF SCIENCE

Here is another Sergeant Major Potts experiment and painting flowers without a paint brush.

Now that the weather is great, I walked outside in my short trousers and combat boots and I spotted some white flowers. So I picked them for an experiment and hopefully the neighbors didn't mind my taking them.

What you will need:

White flowers work the best, hopefully from your own yard, food coloring, cup of water.

1. Add about 20-30 drops of food coloring to a cup of water. In this case, the more food coloring the better!
2. Before placing the flower in the cup of water, have an adult trim the stem of each flower at an angle to create a fresh cut stem. For cut flowers, it is important for the stem tubes to be fresh. If air gets in the tube no water can move up the stem.
3. Place the freshly cut flower in the cup containing the colored water and you may need to wait 24 hours to see results from this experiment.

Now that you have mastered this experiment, try splitting a stem down the middle and place one half stem in one cup of colored water and the other half in a cup of another color. What do you think will happen?

Most plants "drink" water through their roots. The water travels up the stem of the plant into the leaves and flowers where it makes food. When a flower is cut, it no longer has its roots, but the stem of the flower still "drinks" the water.
The process at work here is called "capillary action." Plants have tiny little tubules going up their stems. Think about when eating a piece of celery and those "strings" that get caught in your teeth are what I'm talking about. Water naturally attract itself to these tubes through "adhesion" and gets pulled upward. This is how even the mightiest trees bring water up from their roots to the tallest of their branches.
This is Sergeant Major Potts signing off until next time.

Sergeant Major Potts says. "Keep an eye out for Ducks."

SERGEANT MAJOR POTTS WORLD OF SCIENCE

Potts journal, July 1906. Prepare for another amazing adventure and science experiment with Sgt. Major Potts and the world of science presenting magical things that you can do with milk. Potts was attempting to capture the evil Cornelius Snod who was running through the spice markets of Istanbul. Snod was about to travel to Italy to steal the ancient Phidias head of ivory. Losing Snod around a corner, Potts came to Three Fingered Fred. His other hand was called "one and a thumb." Fred worked at a saw mill. Fred told Potts that Snod went that way, pointing with his fist North and Potts searched on and realized Fred was pointing with his finger and thumb the wrong way and lost sight of Snod. Potts called the museum and asked if they had milk. Milk?????

Perform this experiment and learn how Potts averted Snod from stealing the real head of ivory.

What you will need:
One cup of milk, a sauce pan, and white vinegar.
Place one cup of whole milk in a sauce pan and heat to almost boiling. Next, pour 4 teaspoons of white vinegar into the milk and stir for 3 minutes. Once the mixture cools, pour the milk contents into a strainer and notice that you produced plastic milk. Plastic milk?
Actually, the acidic vinegar makes the milk's fat proteins come together, or coagulate, forming a blob called casein. If you knead the casein blob, it will become clay like and you can make molded figures. This is also the first step to make cheese and cottage cheese.
Placing your figures on paper and in several days the casein will harden. When Potts called the museum he told them to make the plastic milk and to form it in the shape of the head of ivory so Snod will steal milk instead of the priceless ancient museum piece.
In the early 1900's milk was used to make buttons, beads, and other jewelry.

SERGEANT MAJOR POTTS WORLD OF SCIENCE

This experiment will seem simple, every day, and ho hum. But hold onto your hats, the wall, your ears, or anything else to keep you from falling over because you won't believe all of the applications of the words **SURFACE TENSION** that you will discover.

Ok, here we go. You will need a soup bowl, a piece of paper towel, two paper clips, and water.

1. Drop the first paperclip in the bowl of water. What happened?
2. Tear off a piece of paper towel that is larger than the second paperclip.
3. Place the paper towel on top of the water and gently place the second paperclip on the piece of paper towel.
4. Wait a few seconds and slowly push the paper towel down in the water. What happened to the paperclip? Did it float?

The paperclip floated because water particles are attached to each other in all directions, making them "stick" together. Because there are no water particles above them, the water particles at the surface "stick" only to particles next to and below them. This is called Surface Tension.

Now for amazing, beyond belief, applications of **SURFACE TENSION**.
SURFACE TENSION is responsible for the shape of liquid droplets.

SURFACE TENSION provides the necessary wall tension for the formation of soap bubbles. The molecules hug each other and wrap around the air inside producing a spherical shapes.

Tent materials are somewhat rainproof in that the **SURFACE TENSION** of water will bead up over the finely woven tent material. But if you touch the tent material with your finger, you break the **SURFACE TENSION** and water will drip through.

Small insects such as the water strider can walk on water because their weight is not enough to penetrate the surface. They can do this because of …. Why??? **SURFACE TENSION**.

This is Sergeant Major Potts saying be aware. Science is everywhere.

SERGEANT MAJOR POTTS WORLD OF SCIENCE

Potts journal August 1907. Sgt. Major Potts wonders of science and the magical spaghetti noodles. These experiments may amaze, astound, shock. Hey! Someone may pass out when you demonstrate these amazing experiments to thousands. Maybe hundreds. Possibly 50. More than likely 4 or 5.

Sgt. Major Potts had a map leading to the lost Peacock Throne of India's Shah Nader from the 16th century. The gold gilded throne covered with thousands of diamonds, rubies and emeralds has been lost for hundreds of years and Potts has spent years attempting to locate it and to return it to India's museum of antiquities. He put the map in his wallet.

Potts just got back to his hotel in Calcutta India. He put his wallet on the dresser and went to sleep. He awoke to see a shadow go past his bed and head to the open window. Potts grabbed his Z35TWZ regulation flashlight and shined the light on a young boy handing the wallet to an older boy outside of the window. As Potts caught the young boy the second boy disappeared into the darkness with the wallet.

"Gad Zooks", Potts proclaimed. "What super dooper science experiment can I produce to get my wallet back?" He led the boy to the hotel's kitchen. He poured every day and not unusual water into a water pitcher. He then added two contents into the water along with dry spaghetti noodles. Potts then waved his hands over the pitcher and said the magic words, "Shesam-gismo-neato-jet" and the spaghetti noodles began to dance in the water. The young boy looked at this

SERGEANT MAJOR POTTS WORLD OF SCIENCE

amazing, beyond belief spectacle. Potts told the boy to bring back the wallet and he will tell the boy how to produce the magic.

The boy was released and soon brought back the now empty wallet. Potts opened the wallet and produced the map from a secret compartment in his wallet.

Now here is the secret to the dancing spaghetti noodles that you can produce.

Fill a clear water pitcher with every day and not unusual water. Add 3 tablespoons of Sodium bicarbonate (baking soda) and stir the contents. Next pour in one cup of vinegar. Now break a handful of dry spaghetti noodles, one inch or smaller, and place in the water. Wait a few moments and then you say the magic words. Mixing baking soda and vinegar produces carbon dioxide gas. Since carbon dioxide bubbles are less dense that water, they float to the top of water. The bubbles adhere to the spaghetti noodles making them float to the top. The bubbles pop and the spaghetti sink to the bottom of the pitcher. If the noodles don't perform, add more contents to the water. Is this magic? No way! It is science!

This is Sgt. Major Potts world of science signing off until next time. Pip, Pip

SERGEANT MAJOR POTTS WORLD OF SCIENCE

S. M. Potts would like to remind you about duck season. If you see a snow ball coming your way, "DUCK"

Potts journal May 1907. On to more mystifying wonders of Science and the famous balloon rocket.

Sergeant Major Potts was summoned to a Roman archeological dig. There they found a ring that was thought to be the opal ring aged around 30 BC, that Mark Anthony wanted to give to Cleopatra. A Roman senator named Nonius owned the ring and wouldn't give it to Mark Anthony. The ring was never seen again, until now.

Potts was to deliver the ring to the Roman museum but the evil Cornelius Snod wasn't far behind Potts to snatch the famous ring. Potts saw Snod, headed into a building, and flew up a flight of stairs with Snod close behind. On the tenth floor, Potts rushed into a room and locked the door. With Snod pounding on the door, Potts called on his military walkie-talkie to the museum that just happened to be next to the building Potts was in. Looking out of the window, Potts could see the museum director looking at Potts from the museum's window. "Crum Butter", Potts exclaimed. "What to do now?"

Looking in his pocket he pulled out balloons. He also pulled out a straw. He thought, "If only I had some string." He looked down at one of his socks with a loose thread and began pulling and pulling the thread. He tied one end of the thread to a pencil and with a rubber band shot the pencil attached to the thread over to the museum. The museum director held onto the end of thread. Potts blew up the balloon and taped the straw on top of the balloon along with the ring. He put the thread through the straw and held onto the other end of the string and let go of the balloon that traveled to the director. The ring was saved.

Here's a simple and fun science experiment that can be used to teach you about "Action and Reaction". The force from the air moving in one direction propels the balloon in the other direction, much like a rocket. For every action there is an equal and opposite reaction.
First you need: Size 9 or larger balloon, 9 or 10 feet of string, such as kite string (and not from your socks), scotch tape and a straw, around 4 inches long.

SERGEANT MAJOR POTTS WORLD OF SCIENCE

Thread the string through the straw with the other hand, unless you have three, (use your judgment), attach the straw in the center of the balloon with the two pieces of tape.
Tie one end of the string to a chair, a door knob, or find a museum director to hold onto it.
Hold onto the other end of the string so the string is tight.
Blow up the balloon then release it and watch it rocket across the room.

This is S. M. Potts signing off until next time.

SERGEANT MAJOR POTTS WORLD OF SCIENCE

Potts journal September 1907.

I ate some fish sticks that I put on the barbeque and I thought that they tasted like an old tire from a combat Jeep. How do I know what a Jeep tire tastes like? Here is a taste of science fun.

Have you heard someone say, "Dinner smells so good you can almost taste it?" Flavor is taste, smell, texture, and temperature. Taste can be fooled by smells and texture such as a fake apple pie. You can believe you are eating an apple pie after baking Ritz crackers, sugar, lemon juice, butter and cinnamon to make the pie. Information goes to your brain that decides since it feels, tastes, and smells like apple pie, it must be apple pie.

Do you believe that apples, onions, and potatoes could have the same taste? For an experiment close your eyes, pinch you nose and taste a piece of each. The conclusion will probably be that you cannot taste the difference, since the nose plays a big part of tasting food. If you have a cold, is the food you eat tasteless?

A few more fun food facts.
How does popcorn pop? Kernels of popcorn look dry, but each kernel of popcorn has a tiny amount of water inside. When heated, the water molecules turn to steam causing 135 pounds of pressure per square inch. As a kernel explodes, the gooey soft starch molecules release from the kernel and cools immediately, forming the white fluffy popcorn.
Cabbage is 91% water.
Butterflies taste with their feet
Honey is the only edible food for humans that will never go bad.
Cherries are a member of the rose family.
Many mass-produced ice creams have seaweed in them.
Frank mars invented the Snickers chocolate bar. He named it Snickers after his favorite horse.
The Popsicle was invented by an 11-year-old. One cold winter he left his soda water drink, with a stick in it, overnight on his porch.

This is the Sergeant Major Potts World of Science signing off until next time.

SERGEANT MAJOR POTTS WORLD OF SCIENCE

Well, the anticipation is over. Here is another Sgt. Major Potts science secret from the tombs of the pharos and the mystical, magical potato.

What you will need:

One tall glass, 3/4 cup of sugar, and a one inch piece of raw potato.

Fill the glass half full of water. Add 3/4 cup of sugar to the water and stir until the sugar is dissolved. Now place the piece of potato in the glass. Is the potato floating?

Ok, now <u>SLOWLY</u> pour regular and everyday water next to the inside wall of the glass with sugar water to fill the glass. Is the potato floating or sinking? If someone else is watching, this is when you say the magic words, "shisam-gismo-neato-jet."

The sugar water is denser than regular water. I don't mean dense as in not going to school. This dense is like an elevator with 20 people in it is denser than 2 people in an elevator. So the heavy sugar molecules mixed with the water molecules are denser than the potato. So it doesn't sink. But the potato is denser than regular water so it sinks. The potato should remain suspended in the middle of the glass.

Magic? No Way! It's science.

Still not satisfied? If you have a can of regular soda and a can of diet soda then you can do this next amazing, beyond belief experiment. First, fill a pitcher with regular and everyday water. Next place the regular can of soda in the pitcher of regular and everyday water. Did it sink or float? Take that can out and place the diet soda in the pitcher of regular and everyday water. Did that can sink or float?

The science secret is on the side of the regular can of soda. Read how many grams of sugar are in the content? 39 grams of sugar equals 18 packets of sugar. The diet soda has no sugar and is less dense than the heavy, dense, regular soda can.

So if you drink one can of regular soda every day for one year, you will have drunk 32 pounds of sugar. At that point you surely will not float.

This is Sgt. Major Potts signing off until next time saying Pip, Pip…

SERGEANT MAJOR POTTS WORLD OF SCIENCE

To proceed with this experiment you must order the Happy/Sad Ball Kit, SB3394M, from Nasco Scientific. The two balls inside may amaze, astound, shock. Hey! Some-one may even pass out when you demonstrate the amazing bouncing ball. I say ball because you always keep one hidden in one hand. With the other hand, unless you have 3, (use your judgment), you bounce the ball (the one that bounces), and hide the no bounce ball in the other hand. You will then say with an authoritative voice, "This ball has energy." "It has potential energy." "After it smashes to the ground, it then springs up like a spring". "That energy of motion is kinetic energy." After you say that to everyone, without anyone knowing, hand the other ball, (the one that doesn't bounce), to someone and ask them if they have potential energy because you need potential energy to bounce the ball. When they can't bounce the ball, take the ball back and without them knowing bounce the bounce ball. Be ready. That could be the moment that someone may pass out. Sergeant Major Potts would be proud of your dedication to passing on the secrets of the Egyptian tomb magic.

Warning: The magic balls are very small. They can be lost for good if not kept in a container. They are also dangerous if are in contact with dogs. They tend to chew and swallow small balls and I will bet that they will not bounce when it comes out the other end.

ONE MORE AMAZING SPECTICLE

This experiment is called "The famous Baggie Race" that must be performed in the kitchen sink or outdoors. To do this experiment you will first need two sandwich size zip lock baggies, vinegar, baking soda, and two squares of toilet paper. Place 1 table spoon of baking soda in the middle of each of the TP squares and fold or twist the tissues shut. Pour ½ cup of vinegar in each of the bags. Now for the fun part. Quickly drop the tissue of baking soda in each bag and close them up tight. This is when an amazing scientific spectacle will happen. The baking soda and vinegar mix creating a chemical reaction producing carbon dioxide gas that will expand the baggies. Who's baggie bust first?

SERGEANT MAJOR POTTS WORLD OF SCIENCE

On to Potts incredible, extraordinary experiments using balloons
When it is cold outside and family and friends are together, here are some indoor scientific activities and challenges that you can present using balloons. With these experiments you will need every day and not unusual balloons. Use balloons preferably size 9 or larger.

1. Bouncing balloons, sports, and engineering: For this challenge you will need balloons, scotch tape, and coins. Have competitive groups or individuals compete by inflating their balloon, tie it shut, and tape coins on it. Experiment by placing the coins in different locations to make the balloon bounce. Then compete to see whose balloon bounces the most times.

2. Before inflating a balloon, place a coin inside of the balloon, then inflate the balloon and tie it shut. Twirl the balloon so the coin will travel around inside of the balloon. This is called centripetal force that generates an inward pull on the coin against the inner wall of the balloon. Now place a hex nut in another balloon. Inflate that balloon and tie it shut. Again twirl the balloon and be surprised by the vibrating sound of the centripetal forced nut. The six sides of the hex nut generate vibrating sounds along the inner wall of the balloon. The sound should drive cats and moms crazy.

3. To present a magic trick, inflate a balloon and tie it shut. Place ½ inch long pieces of scotch tape at different locations on the balloon. Next, twist one or several needles through the locations where the scotch tape is located. Then you say the magic words, "shisam-gismo-neato-jet". The balloon should not pop. The tape holds the balloon together keeping it from tearing and popping.

4. Blow up two balloons the same size and tie them shut. Put one balloon in the freezer and leave the other one out. After 24 hours, take the balloon out of the freezer. Compare the two balloons. You will see that the balloon from the freezer is now smaller. Cold air molecules take up less space than warm air molecules. As the air in the balloon from the freezer warms, the balloon will once again become the same size as the other balloon.
We hope you enjoyed experimenting with the incredible, extraordinary effects using balloons. This is Sgt. Major Potts World of Science signing off until next time.

SERGEANT MAJOR POTTS WORLD OF SCIENCE

Potts journal July 1907. More exploits to the incredible, extraordinary, awesome Sergeant Major Potts World of Science and the rattling magnets. Don't hold your breath because you will need all the oxygen you can get for this one.

Potts was frantically putting a tarp over his jeep in pouring down rain when he heard his telephone ring inside. He rushed into the barracks to receive a call from his close friend, Chidiebube, in Serengeti, Africa. The phone reception was very bad and all that Potts could hear was that he must desperately go quickly to Africa because of lox and donta. "Lox and donta?" Potts questioned. He scratched his head and chomped on his cigar. "Lox is bagel and fish and donta must be Oliver Donta the great explorer of pastoral Neolithic stone age diggings in Africa." And he must like to eat lox. "Gad zukes", Potts exclaimed. "Oliver Donta must be in danger and needs my help."

Once again Potts got on a Clipper airplane and headed to Serengeti Africa.
Soon after Potts got off of the plane in Serengeti, two tall guys in black trench coats grabbed him and threw him in a car and drove him to a two-seat biplane. Potts recognized the pilot as the notorious, evil, elephant tusk-poaching Cornelius Snod. Snod said, "I heard you were coming to stop my notorious, evil, elephant tusk poaching so I have come to stop you from stopping me." "Crum butter," Potts declared. " Chidiebube wasn't saying lox and Oliver Donta." "He was trying to say loxodonta." "That means elephant in Latin and Snod isn't a nice guy." The plane took off with Cornelius Snod and Potts. Potts searched in all of his pockets for a diabolical mystifying Egyptian science experiment to get him out of this situation. From the top inside pocket of his jacket he pulled out two torpedo-shaped magnets, yelled "snake" and threw them into pilot Snod's lap. The magnets made a vibrating sound like rattlesnakes. Snod jumped so high in his seat that he fell out of the plane. It was Lucky for Snod that he had a parachute. This gave Potts the extra time to fly back to the police authorities to reveal Snod's bad deeds and inform them of his location. Once again Potts saved the day.

SERGEANT MAJOR POTTS WORLD OF SCIENCE

What is a torpedo magnet you say? First, what is a magnet? It is a device that attracts iron from its magnetic field. What is a magnetic field? Boy you ask a lot of questions. There are invisible moving particles in an iron magnet that move from the top to the bottom and grab onto metal and don't want to let go until you pull them apart. The torpedo magnets grab each other but their shape makes them vibrate to produce an unusual sound. Where can you find such magnets? Well, go to your computer and go to "Amazon" and type in Torpedo Magnets. This is S. M. Potts saying, fly straight, keep on course, and don't crash.

SERGEANT MAJOR POTTS WORLD OF SCIENCE

More amazing, beyond belief experiments from Sergeant Major Potts and follow these experiments for the Potts, puzzling, paper presentation.
Ok scientists, here are three experiments using paper.

1. The first experiment you will tell someone that you can climb through a hole in this piece of paper. If they look at you cross eyed, just follow the 3 steps and let them be amazed when you unfold the paper that you cut and step through the opening. Any size of paper will do. 1. Fold the paper in half and cut from the fold but not all the way to the other side.
2. Cut at the crease from the first cut to the second cut but don't cut the ends.
3. Now cut as shown, unfold, and now you have a hole in the paper that you can climb though.

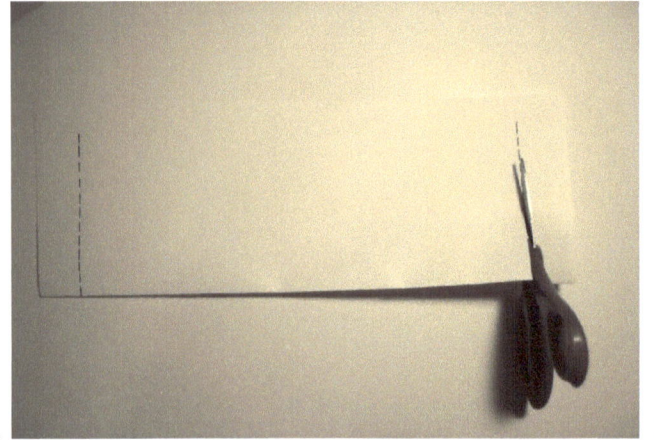
1.
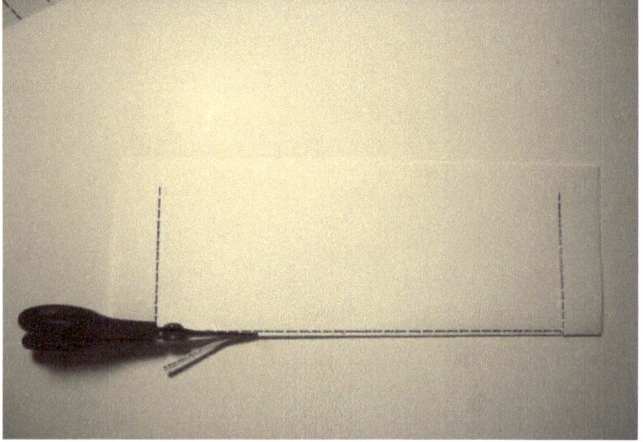
2.
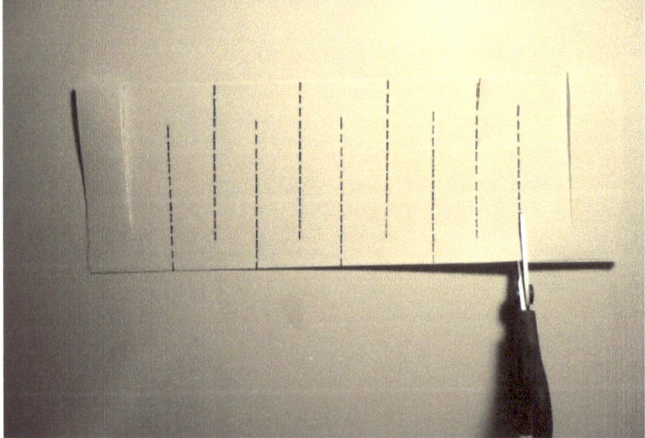
3.

4. With a thumbtack or pin, push a small hole in an index card or paper. Look at the small letters in a book through the paper pinhole. If you are nearsighted and wear glasses, take off your glasses. You will see both near and far away objects more clearly through the pinhole than you do if you look at them without the aid of the pinhole.

SERGEANT MAJOR POTTS WORLD OF SCIENCE

How this works: When light rays reflected from the book and pass through the tiny hole in the index card, the rays are pushed more closely together. The light rays become focused by a small part of the eye's lens on the retina (the light-sensitive back of the eye). This focusing effect causes blurred images to become clearer. More science fun: Change the size of the pinhole. How large can the hole be before it no longer clarifies images?

And with the third experiment take two paper plates, cut two exact same sizes around the edges as shown below. Don't forget to cut an angle on the edges as shown. Now ask someone which one is bigger. If they say B, place B above A and ask them again to tell you which one is bigger.

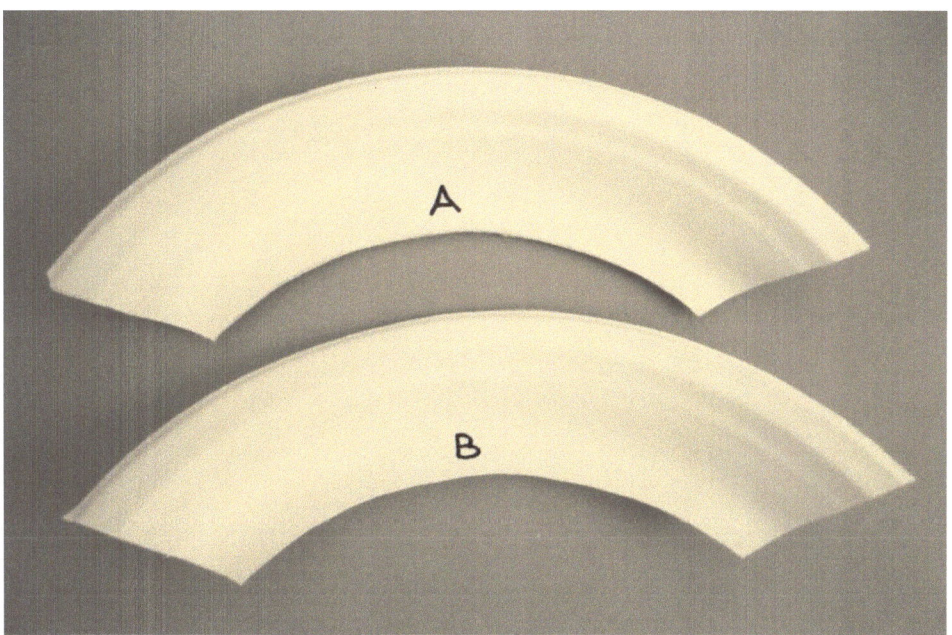

Amazing? Yes! Is it magic? NO WAY!!! It is science. Remember, science is everywhere.

SERGEANT MAJOR POTTS WORLD OF SCIENCE

Cheers Science enthusiasts from the Sergeant Major Potts World of Science. It has been an honor to have you as copilot in the exploits of the World of Science. You have done us proud passing on the discoveries of science to the minds of thousands. Maybe hundreds. Possibly 50. More than likely 10 to 15.

In this journey of science, we out foxed poison dart pigmies, bamboozled Jajari soldiers, dazzled and hornswoggled pirate looters.
Potts had motivation to search for mysteries of science. He wasn't too fond of fishy, smelly, sharp harpoons so he used imagination from what he learned that opened new doors of discovery.

Now it is time for you to find new discoveries using motivation and imagination.
It was once said, "Motivation doesn't last." Neither does taking a bath." "That is why they are recommended daily." (Zig Ziglar)

This is the Sergeant Major Potts World of Science saying pip, pip, cheerio and signing off.

www.ingramcontent.com/pod-product-compliance
Lightning Source LLC
Chambersburg PA
CBHW050427180526
45159CB00005B/2436